The Wildfire

The unexpected event that took Wharton State Forest

By

Madelyn Wilson

Table Of Contents

- Batsto Village Attractions
- What Is The Best Way To Get To Batsto Village?
- Nearby Attractions

Conclusion

Introduction

Wildfires can occur at any time or in any area, and they are usually sparked by human action or natural events like lightning. Half of the wildfires that have been reported have no identified cause.

Extremely dry circumstances, such as drought, and strong winds make wildfires more likely. Wildfires can affect transportation, communications, power and gas supplies, and water delivery. They also lead to a reduction in air quality and the loss of property, crops, resources, animals, and people.

Between 1998 and 2017, wildfires and volcanic eruptions killed 2400 people worldwide due to suffocation, injuries, and burns, however, the magnitude and frequency of wildfires are growing as a result of climate change. Ecosystems are drying out as a result of climate change, increasing the danger of wildfires. Wildfires affect weather and climate by releasing massive amounts of carbon dioxide, carbon monoxide, and fine particulate matter into the atmosphere. As a result, air pollution can lead to a number of health issues, including respiratory and cardiovascular issues. Wildfires also have a substantial impact on mental health and psychological well-being, and the damage caused by a wildfire is dependent on its magnitude, extent, and other circumstances.

Chapter 1
Wildfire

A wildfire is an uncontrolled fire that burns in the vegetation of the wilderness, usually in rural locations. Wildfires have been burning for hundreds of millions of years in forests, grasslands, savannas, and other habitats. They aren't limited to a single continent or ecosystem.

Extremely dry circumstances, such as drought, and strong winds make wildfires more likely. Wildfires can affect transportation, communications, power and gas supplies, and water delivery. They also lead to a reduction in air quality and the loss of property, crops, resources, animals, and people.

Wildfires can start both below and above the ground-level vegetation. Ground fires are most likely to start in soil that is rich in organic components, such as plant roots, that can fuel the flames. Ground fires can smoulder for months, even years, waiting for the right conditions to turn them into a surface or crown fire. Surface fires, on the other hand, are started by dead or dry vegetation that is laying or growing just above ground level. Parched grass and fallen leaves are common sources of fuel for surface fires. Crown fires start in the leaves and canopies of trees and shrubs.

Different sorts of wildfires can inflict some areas, such as the mixed conifer forests of California's Sierra Nevada mountain range. Forest fires in the Sierra Nevada can contain both crown and surface areas.

A wildfire can be started by a natural occurrence, such as a lightning strike, or by a human-made spark. Weather factors, on the other hand, frequently impact the amount to which a wildfire spreads. Wind, heat, and a lack of rain may all cause trees, bushes, fallen leaves, and branches to dry out, making them ideal firewood. Flames burn faster uphill than downhill, therefore topography is important. If wildfires burn too close to people's homes, they can become dangerous and even deadly.

The 2018 Camp Fire in Butte County, California, for example, destroyed practically the entire town of Paradise, killing 86 people. Wildfires, however, are necessary for the survival of several plant species. Some tree cones, for example, require heat to open and release their seeds; chaparral species such as manzanita, chamise (Adenostoma fasciculatum), and scrub oak (Quercus berberidifolia) all require fire to germinate their seeds.

The leaves of these plants contain a flammable substance that helps the plants reproduce by feeding fire. Plants like these rely on wildfires to complete their life cycle regularly. Some plants need to be burned every few years, while others just need to be burned a few times every century to survive. Wildfires also contribute to the health of ecosystems. They can eliminate insects and diseases that are harmful to trees. Fires can clear scrub and underbrush, allowing new grasses, herbs, and shrubs to emerge, providing food and habitat for animals and birds. Flames can sweep up waste and underbrush on the forest floor, add nutrients to the soil, and open up space for sunlight to reach the ground at a low intensity. Smaller plants will benefit from the extra light, while larger trees will have more room to develop and thrive.

While wildfires are necessary and beneficial to many plants and animals, climate change has made some ecosystems more vulnerable to fires, particularly in the southwest United States. Drought has worsened as a result of rising temperatures, and forests have dried up. The historical tradition of putting out all fires has resulted in an unnatural buildup of vegetation and garbage, which can feed larger, more severe fires.

A single spark or even the heat of the sun can ignite an inferno in seconds. The wildfire spreads swiftly, devouring the thick, dried-out foliage as well as nearly everything else in its path. What was once a forest has now turned into a virtual explosive bomb of untapped energy. The wildfire engulfs thousands of acres of surrounding land in an apparent instantaneous burst, endangering the homes and lives of many in the area.

Every year, the United States burns an average of 5 million acres, causing millions of dollars in damage.

Once started, a fire can spread fast up to 14.29 miles per hour (23 kilometres per hour), destroying everything in its path. A fire may take on a life of its own as it burns through grass and trees, searching out new ways to stay alive and even spawning smaller fires by tossing embers kilometres away. In this post, we'll look at wildfires and how they originate, grow, and die.

A single spark from a train car's tire striking the track on a hot summer day, when drought conditions are at their worst, can start a roaring wildfire. Fires can start naturally, sparked by the sun's heat or a lightning strike. The bulk of wildfires, on the other hand, are caused by human negligence.

The following are some of the most common causes of wildfires:

- Arson
- Campfires
- Putting out lighted cigarettes
- Debris that has been improperly burned
- Matches or fireworks that are being played with.
- Prescribed fires

Every object has a temperature at which it will combust. The flash point of a material is the temperature at which it melts. The flash point of wood is 572 degrees Fahrenheit (300 C). When wood is burned to this temperature, hydrocarbon gases are released, which combine with oxygen in the air and combust, resulting in a fire.

For ignition and combustion to occur, three elements are required. A fire needs fuel to burn, oxygen to breathe, and a heat source to bring the fuel up to ignition temperature. The fire triangle is made up of three elements: heat, oxygen, and fuel. When firefighters are attempting to put out a fire, they frequently refer to the fire triangle. The notion is that if they can remove any one of the triangle's pillars, they will be able to control and eventually extinguish the fire.

Various factors impact how a fire spreads after combustion occurs and it begins to burn. Fuel, weather, and topography are three of these three factors. A fire might rapidly fizzle or turn into a blazing conflagration that scorches thousands of acres depending on these circumstances.

Fuel Load

Wildfires spread based on the type and quantity of fuel available around them. Everything from trees, underbrush, and dry grassy fields to houses can be used as fuel. The fuel load refers to the amount of flammable material that surrounds a fire. The amount of available fuel per unit area, generally tons per acre, is used to calculate the fuel load.

A limited fuel load will produce a low-intensity fire to burn and spread slowly. The fire will burn more intensely if there is a lot of fuel, leading it to spread faster. The faster it heats the materials in its vicinity, the more likely they are to ignite. The dryness of the fuel can also influence how the fire behaves. When the fuel is extremely dry, it is used significantly more quickly, resulting in a much more difficult-to-control fire.

The following are the basic qualities of fuel that determine how it causes a fire:

- Dimensions and form
- Arrangement
- Moisture level

Small fuel sources, often known as flashy fuels, burn faster than massive logs or stumps, such as dried grass, pine needles, dry leaves, twigs, and other dead vegetation. Different fuel compounds take longer to ignite chemically than others. However, in a wildfire, if the majority of the fuel is formed from the same type of material, the ratio of the fuel's total surface area to its volume is the most important factor in igniting time. The surface area of a twig is not much larger than its volume, therefore it ignites quickly. A tree, on the other hand, has a considerably smaller surface area than its volume, therefore it takes longer to heat up before igniting.

The material just beyond the flames dries out as the fire proceeds; heat and smoke approaching potential fuel cause the fuel's moisture to evaporate. When the fire approaches the fuel, it will be easier to ignite.

Fuels that are somewhat spaced out will likewise dry out faster than fuels that are packed firmly together because the thinned-out fuel has more oxygen accessible. Fuels that are packed more firmly retain more moisture, which absorbs the heat from the fire.

The Role of Weather in Wildfires

A wildfire's genesis, growth, and death are all influenced by the weather. Drought creates ideal conditions for wildfires, and winds promote their spread — the weather can encourage the fire to spread faster and consume more area. It can also make the task of putting out the fire more difficult. Wildfires can be influenced by three weather factors:

- Temperature
- Wind
- Moisture

Because heat is one of the three pillars of the fire triangle, temperature influences the occurrence of wildfires. The sun heats and dries potential fuels by radiating heat through the ground's sticks, trees, and underbrush. Warmer temperatures encourage fuels to ignite and burn more quickly, hastening the development of a wildfire. As a result, wildfires are more likely to rage in the afternoon, when temperatures are at their peak.

The wind has the greatest influence on the behaviour of a wildfire. It is also the most unpredictably unexpected factor. Winds contribute oxygen to the fire, dry out potential fuel, and speed up the spread of the fire across the terrain.

A computer model that depicts small-scale wind movement was created by a scientist and that model has been updated since 1991 to add wildfire characteristics like fuel and heat transfer between fires and the atmosphere.

According to research, not only does wind influence how a fire develops, but fires can also develop wind patterns. They can feed back into how the fire spreads when the fire develops its weather patterns. Winds known as fire whirls can be generated by large, severe wildfires. Fire whirls, which resemble tornadoes, are caused by the vortices formed by the heat of the fire. Fire whirls are created when these vortices are tilted from horizontal to vertical. Fire whirls have been reported to propel flaming logs and burning debris a long way.

There's another approach to changing the vorticity's direction. That is, it can be named without exploding into fire whirls, and can instead be hurled forward into hairpin vortices or forward bursts. You see fires licking up hillsides because these are extremely prevalent in crown fires [fires at the top of trees]. Forward bursts can be 20 meters wide (66 feet) and shoot out 100 meters (328 feet) at 100 miles per hour (161 kph). These explosions burn the area and cause the fire to spread.

The faster the fire spreads, the stronger the wind blows. The fire creates its winds, which can be up to ten times faster than the surrounding wind. It can even throw embers into the air and start new flames, a phenomenon known as spotting. Wind can also shift the fire's course, and gusts can lift the fire into the trees, causing a crown fire.

While the wind can aid in the spread of a fire, moisture acts against it. Moisture, in the form of humidity and precipitation, can help to lower the severity of a fire. Because moisture absorbs the heat of the fire, potential fuels with high moisture levels can be difficult to ignite. Wildfires are more likely to start when the humidity is low, meaning there isn't much water vapour in the air. The less probable the gasoline is to dry out and ignite, the higher the humidity.

Precipitation has a direct impact on fire prevention since moisture can reduce the odds of a wildfire beginning. The moisture in the air is released in the form of rain when it becomes saturated with moisture. Rain and other precipitation increase the moisture content of fuels, preventing any possible wildfires from igniting.

Fire On The Mountain

The lay of the terrain, or topography, is the third major factor influencing wildfire behaviour. Unlike fuel and weather, terrain, which remains practically unchanging, can help or hinder wildfire progression. The slope is the most essential component in topography when it comes to wildfires.

Unlike humans, fires like to travel uphill rather than downhill. The higher the slope, the faster the fire moves. Fires usually spread uphill in the direction of the prevailing wind. Furthermore, because the smoke and heat are rising the hill, the fire might warm the fuel further up the hill. In contrast, once a fire reaches the top of a hill, it must work hard to get back down since downhill fuel does not preheat as efficiently as uphill fuel.

Fires moving more slowly uphill are the exception rather than the rule. Winds can make it difficult for a fire to advance up a slope. It depends on which way the wind is blowing. For example, in Australia, the wind was blowing down the mountainside, pushing the fire away from the hill until a front passed through. After that, it was all uphill.

Aside from the harm that flames inflict as they burn, they can also leave behind disastrous problems that may not be felt for months after the fire has been extinguished. When fires burn down all of the vegetation on a hill or mountain, it weakens the organic content in the soil and makes it difficult for water to penetrate. One issue that arises as a result of this is extremely severe erosion, which can result in debris flows.

A wildfire that burnt about 2,000 acres of timber and underbrush on the steep slopes of Storm King Mountain near Glenwood Springs, Colorado, in July 1994 was an example of this.it rained strongly two months after the fire created debris flows that washed tons of mud, rock, and other debris onto a three-mile length of Interstate 70. 30 cars were enveloped in the debris flows, two of which were swept into the Colorado River.

While many people think of wildfires as destructive, many of them are beneficial. Some wildfires burn a forest's underbrush, preventing a larger fire from forming if the brush is allowed to build for a long time. Plant growth can also be aided by wildfires because they reduce disease transmission, release nutrients from burned plants into the ground, and encourage new growth.

Battling The Blaze

Imagine being inside an oven, wearing heavy gear, with smoke filling your lungs, and you have a rough idea of what it's like to battle a roaring wildfire. Thousands of firemen risk their lives every year to fight ferocious fires. There are two types of elite ground-based firefighters:

1. Hotshots - These highly trained firefighters work in 20-person teams to create a firebreak around the fire to prevent it from spreading. A firebreak is a section of land that has been cleared of all potential fire fuel. The United States Forest Service employs hotshots.

2. Smokejumpers - These firemen are paratroopers who jump out of planes to reach small fires in distant locations. Their mission is to put out little fires before they grow into larger ones. Once they've landed, Smokejumpers use the same firefighting procedures as the Hotshots. Only a few hundred smokejumpers work in the United States, all of

whom are hired by the Bureau of Land Management (BLM) or the United States Forest Service.

Ground teams may utilize backfires in addition to establishing firebreaks and dousing the fire with water and fire retardant. Backfires are fires caused by ground crews as they approach a raging wildfire. Setting a backfire has the purpose of consuming any potential fuel in the path of a raging wildfire.

3. Smokejumpers -Smokejumpers, and other support personnel fight on the ground, they receive a great deal of help from the air. Thousands of gallons of water and fire retardant are frequently dropped onto fires by air tankers. The red substance you see being thrown from planes and helicopters is a phosphate fertilizer-based chemical retardant that helps to slow and cool down the fire.

Helicopters can also be employed to attack the flames from the air. These planes, which are equipped with buckets that can contain hundreds of gallons of water, fly over the fire and drop water bombs. Helicopters are also useful for getting firefighters to and from a fire.

Wildfires are tremendous natural forces that can burn for as long as fuel, oxygen, and heat are available. To prevent further damage, firefighters must destroy one, if not all three, sides of the fire triangle.

Chapter 2
Forest Fire

Forest fire is an uncontrolled fire that starts in vegetation that is taller than 1.8 meters (6 feet). These fires can swiftly grow to the magnitude of a massive conflagration, and they're frequently sparked by surrounding surface and ground fires. A major forest fire may crown, or quickly spread through the highest branches of trees, before spreading to the undergrowth or forest floor. As a result, forest fires frequently experience strong blowups, which might resemble a firestorm.

Though forest fire is frequently thought to be detrimental, several forest ecosystems are specifically fire-adapted, meaning that the species of plants and animals unique to such ecosystems benefit from or to survive and breed, they rely on the incidence of fire.

In a dry forest, lightning strikes are common, and fire can benefit ecosystem health by lowering competition, enriching the soil with ash, and reducing illneses and pests. Some plant species even require fire to germinate their seeds.

Years of fire exclusion and suppression in many regions that have historically seen forest fires, such as wooded parts of the western United States, allowed fuels to collect, altering vegetation communities and contributing to more intense conflagrations when fires do occur. Prescribed fire, in which regions are burned purposefully and under controlled conditions, can help to recover ecosystems and foster conditions that existed before wildfire was eliminated.

The most common threat in woods is a forest fire. Forest fires have been there since the dawn of mankind. They jeopardize not only the forest's richness, but also the entire fauna and flora regime, wreaking havoc on the region's biodiversity, ecology, and environment. When there has been no rain for months, the woodlands become covered in dried senescent leaves and twinges, which could burst into flames if lighted by the least spark. The Himalayan forests, notably in the Garhwal Himalayas, have been burning often in recent summers, resulting in a major loss of vegetative cover in that region.

The Causes Of Forest Fires

Forest fires are triggered by both natural and man-made sources.

1. Natural causes - Many forest fires are started by natural causes such as lightning striking trees and setting them ablaze. Rain, on the other hand, quickly extinguishes such fires while causing minimal damage. The combination of high temperatures and little humidity creates ideal conditions for a fire to ignite.

2. Man-made causes - When a source of ignition, such as a bare flame, cigarette or bidi, electric spark, or any other source of ignition, comes into contact with flammable material, it generates fire.

Forest Fire Classification

Forest fires can be broadly divided into three types:

- Forest fires can be natural or man-made.
- Forest fires are caused by heat generated in the trash and other biomes by people's negligence (human neglect) in the summer.
- Forest fires were deliberately set by residents.

Types Of Forest Fires

The following are the different forms of forest fires:
The following are the causes of forest fires:

1. Surface Fire - A forest fire may burn largely as a surface fire, spreading throughout the forest floor when surface litter (senescent leaves and twigs, dry grasses, etc.) is absorbed by the expanding flames.

2. Underground Fire - Underground fires are low-intensity flames that consume the organic stuff beneath and the surface litter of the forest floor. On top of the mineral soil, a thick layer of organic matter can be found in the densest forests. By eating such things, the fire spreads throughout the area. These fires usually burn for several meters below ground level and spread fully underground.

This sort of fire spreads slowly, making it difficult to identify and control in the majority of situations. They might burn for months, destroying the soil's vegetative cover. Muck fires are another name for this type of fire.

3. Ground Fire- Ground fires start in subsurface organic fuels such as duff layers under forest stands, Arctic tundra or taiga, and swamp or bog organic soils. It's difficult to tell the difference between underground and ground fires. At any time, smouldering subsurface fires could erupt into a ground inferno. This fire eats roots and other material on or beneath the surface, i.e., herbaceous growth on the forest floor as well as a layer of decaying organic debris. Because they can completely kill plants, they are more deadly than surface fires. Because they burn under the surface via smouldering combustion, surface flames are more prone to initiate ground fires.

4. Crown Fire - A crown fire is a fire that starts at the tops of trees and bushes and is typically fueled by a surface fire. Crown fires are especially harmful in coniferous forests because the resinous substance released by burning logs burns rapidly.
If a fire starts on a hill slope and spreads downhill, it will swiftly move uphill as warm air near the slope

rises, carrying flames with it. If a fire begins uphill, it is less likely to spread downhill.

5. Firestorms - Among forest fires, the firestorm, which is an intense fire across a vast region, spreads the fastest. The fire burns, the temperature rises and air rushes in, forcing the flames to spread.
More air causes the flames to spin violently as if they were a tornado. Flames erupt from the base of the fiery twister, and burning embers erupt from the top, igniting smaller fires all around it. Temperatures inside these storms can reach 2,000 degrees Fahrenheit.

Vulnerability

The Himalayan mountain ranges, which are the world's youngest, are the most vulnerable to forest fires. Forest fires are more common in the Western Himalayan forests than in the Eastern Himalayan forests. This is because the woods of the Eastern Himalayas receive a lot of rain. Forest fires have become more frequent and intense as chirr (Pine) forests have expanded across various sections of the Himalayas.

Chapter 3
Wharton State Forest

Wharton State Forest is in the heart of the Pinelands and is New Jersey's biggest single tract of state-owned territory, with well over 120,000 acres. At every turn, there are important natural resources, historic villages, and recreational options. Batsto Village, a historic bog iron and glassmaking industrial hub from 1766 to 1867 that now depicts the agricultural and commercial operations that were here throughout the late 19th century, is also located here (open for self-guided exploration).

Rivers and streams for canoeing, hiking routes (including a major stretch of the Batona Trail), miles of unpaved roads for mountain biking and horseback riding, and several lakes, ponds, and fields for wildlife observation may all be found across Wharton.

South Jersey became a significant industrial centre due to its abundant natural resources. More than 30 of Batsto Village's 19th-century structures remain, including the completely furnished Batsto Mansion, which made iron items for the Revolutionary War. Wharton offers hiking, mountain biking, and horseback riding paths, as well as canoeing, kayaking, and fishing in rivers and lakes. Picnic tables and charcoal grills, two small playgrounds, a short nature trail, a canoe/kayak launch place, restrooms, and changing facilities are also available at the Atsion Recreation Area.

Historic Towns In Wharton State Forest

In what is now Wharton State Forest, a complex of manufacturing centres prospered in the eighteenth and early nineteenth centuries. Iron, glass, lumber, and paper were created in communities like Bulltown, Harrisville, Friendship, and Martha, which were fueled by natural resources. While many of these locations have now vanished, Batsto and Atsion still have buildings and structures that show the agricultural and commercial endeavours that existed during the nineteenth century.

From 1766 through 1867, Batsto Village was a major iron and glass manufacturing hub.

Natural Environments

Several woodland habitats indicative of the Pinelands can be found in the Batsto Natural Area (9,449 acres). Many rare and endangered plants, as well as some species found nowhere else on the planet, can be found in these locations. The Natural Areas Act of 1961 was enacted to protect natural areas. Sign at the entrance to Batsto Village established an exceptional level of protection for ecologically significant areas within the state of New Jersey that owns these lands.

Joseph Wharton

In 1873, a wealthy manufacturer named Joseph Wharton began buying enormous swaths of land in the Pinelands, including the settlements of Batsto and Atsion. On his gentleman's farm, Wharton experimented with agricultural and forest management. Wharton had amassed 96,000 acres by the time he died in 1909. Because of his efforts, this enormous region has stayed undeveloped and has become the heart of Wharton State Forest.

Opportunities For Recreation

Wharton State Forest's natural beauty can be admired at any time of year. The outdoor enthusiast will find miles of trails, gorgeous waterways, and unspoiled nature. Camping is only permitted in a few authorized sites and only with a permit. Swimming is permitted at the Atsion Recreation Area during the summer months. Batsto Village's visitor centre is open all year. There are various recreational possibilities to experience the area's world-class natural resources, including canoeing and kayaking.

Chapter 4
Wharton State Forest Fire

A forest fire raged through Wharton State Forest on Sunday 19th afternoon, driven by dry and breezy conditions. The fire had spread through 600 wooded acres in Washington, Shamong, Hammonton, and Mullica townships by 7 p.m.Six structures at the Paradise Lakes Campground near Hammonton were threatened. Around 10% of the fire had been put out by 7 p.m and smoke from the fire could be visible as far east as Long Beach Island.

Batsto Village in Washington Township was closed, as was the Mullica River Campground in Hammonton, as well as neighbouring trails and boat launches. Six structures in the Paradise Lakes Campground in the forest were threatened, requiring evacuations. Kayak and canoe trips were also halted by Pineland Adventures and Batsto Village was decommissioned.

The Forest Fire Service was called to a blaze that started near the Mullica River in a remote portion of Wharton State Forest. The fire was 100 acres at 4:30 p.m., but it had grown to 600 acres in just three hours.

Relative humidity levels were as low as 25% in Hammonton and 31% in Oswego Lake, Burlington County, near the fire. Conditions were favourable for wildfire spread, with persistent winds as high as 23 mph in Hammonton and 28 mph in Oswego Lake.

From the Atsion Recreation Area to the ancient Batsto Village bog ore smelting site, the Mullica River Campground, Mullica River Trail, and boat launches along the river have been closed until further notice. Visitors were not permitted to enter Batsto Village or any of the surrounding hiking and mountain biking paths. The Paradise Lakes Campground, which was evacuated by personnel, is threatened by six structures.

No further information was provided from the state Department of Environmental Protection, which includes the Forest Fire Service, and any updates were directed to the Twitter and Facebook accounts. The weather in the area was sunny and dry on Sunday, with temperatures in the mid-70s and wind speeds in the mid-to-high teens.

Wharton is the state's largest state park, with 122,800 acres of pine forest, meadows, lakes, and rivers in Burlington and Atlantic counties. Around 7,000 acres of state forest land are damaged or destroyed by spontaneous fires on an annual basis. The smoke was carried to the beaches by the northwest winds. On Twitter, people reported seeing smoke near Atlantic City and Ocean City, around 25 to 30 miles away from the fire's source.

Chapter 5
Left Fork Fire

ST. GEORGE, S.D. — Due to heavy winds and red flag conditions, a wildfire that was supposed to be under control in the Dixie National Forest in early May has rekindled, burning 600-700 acres of forest property.

A wildfire in Kane County was reignited by strong winds around 1 p.m. Saturday. The fire has been driven northeast to the Blubber Creek drainage by the wind, which is just 10% contained. The fire started on May 9 and burned around 97 acres 3.2 miles west of the Podunk Guard Station. The fire was expected to maintain "minimum, creeping behaviour" within a set boundary at the time, but current weather circumstances aggravated the issue.

The fire, which is burning mixed conifer, ponderosa pine, slash piles, and dead and fallen woody material about 10 miles southwest of Bryce Canyon National

Park near Kanab Creek was human-caused. No structures were in danger, and no evacuation orders were been issued. However, some campers voluntarily left the area.

Firefighters also worked in challenging terrain and terrible weather conditions, which hampered their operations. The low relative humidity was accompanied by southwest winds averaging 20 mph with gusts of 35-40 mph.

Due to the weather, aircraft support, including two assigned helicopters, single-engine air tankers, and big air tankers, were been grounded.145 people and four crews were been dispatched, along with seven fire engines, one dozer, and one fire tender. Responders are striving to create an anchor point and, where practicable, build lines.

A Color Country Type 3 Incident Management team also arrived on the ground and also took leadership on Sunday evening. The rapidly developing fire had weakened standing trees, which, together with dead snags, were at risk of falling and striking firefighters.

The fire was being fought by the Forest Service of the United States, the Bureau of Land Management, the Utah Division of Fire, Forestry and State Lands, Bryce Canyon National Park, Zion National Park, and the Garfield County Sheriff's Office. A Fire Management Officer with the Division of Forestry, Fire and State Lands, Cedar Mountain Fire Protection District and Dammeron Valley Fire and Rescue sent paramedics and Emergency Management personnel to offer emergency medical care on the fire line if needed.

Chapter 6
Batsto Village

Batsto Village is a historic site in New Jersey's South Central Pinelands. It is well-known across the country for its beauty as well as its historical significance. Batsto has origins reaching back to 1766.

History Of Batsto Village

Batsto Village is a nationally designated historic site in Wharton State Forest in Southern New Jersey. Throughout hundreds of years of American history, the Village has altered and survived, and evidence of several thousand years of land use may be found. In the Batsto area, archaeological investigations have even uncovered prehistoric life remains.

The Batsto Iron Works, which was built along the Batsto River in 1766, is credited to Charles Read. Batsto possessed the natural resources required to produce iron. Bog ore was mined from the banks of streams and rivers, timber from the forests was used to make charcoal, and water was used to power industries. The Iron Works produced domestic products, such as cooking pots and kettles. Batsto supplied the Continental Army with provisions during the Revolutionary War.

In 1779, manager Joseph Ball purchased Batsto Iron Works. In 1770, a Philadelphia merchant named John Cox acquired a part-owner of the plant, and by 1773, he was the sole owner.

Years of the Richards

In 1784, William Richards, Joseph Ball's uncle, became a substantial shareholder in Batsto Iron Works. Richards' reign at Batsto would be 92 years long. William was ironmaster until 1809 when he resigned; his son Jesse took over until 1854 when he died; Jesse was succeeded by his son Thomas H.

After iron manufacture fell out of favour in the mid-nineteenth century, Batsto became noted for its window glass. The glass firm soon followed suit, and Batsto was forced to file for bankruptcy.

Wharton Years

During the Wharton years, Joseph Wharton, a Philadelphia merchant, bought Batsto at a Masters Sale in 1876, as well as property in the surrounding area. Wharton improved the home as well as many of the community structures. He was also engaged in several forestry and agricultural ventures.

In 1909, Joseph Wharton passed away. The Girard Trust Company in Philadelphia maintained the Wharton holdings in the Pine Barrens from his death until 1954.

State ownership

In the mid-1950s, New Jersey purchased the Wharton assets and began planning for their usage and development. The few persons who remained in the Village residences stayed as long as they wanted; the last house was vacated in 1989.

Batsto Village is now a nationally recognized historic property, with listings on both the New Jersey and National Registers of Historic Places.

Batsto Village Attractions

Journey into the past is Batsto Village's tagline, and that's exactly what you'll do at this charming, rustic resort. Begin your visit at the Visitor's Center, where you can read about the fascinating history of this unique location and plan whatever tours or activities you want to see and do while you're there. Here are a few highlights:

1. Museum: The Batsto Village museum and gallery is a great place to receive a map of the area and learn about the history of the area. This museum, which has numerous exhibits, is located at the Visitor's Center. It houses a variety of permanent and temporary exhibitions about the New Jersey Pinelands, the village's founding, and historical landmarks. Visit the little gift shop to see what unique presents and children's goods are available.

2. Library: This on-site library, which was established during the hamlet's restoration (about 2004), is brimming with literature about the village and its history, as well as a wealth of knowledge about the surrounding Southern New Jersey area.

3. Post Office:This post office is one of the four oldest in the United States and it is a favourite among history aficionados. It's still open today, despite being closed a few times throughout the years. It's a true step back in time because all stamps are still hand-cancelled.

4. Structures on the property: The Visitor's Center has information on various historic structures on the property. Cottages, a gristmill, a nature centre, a sawmill, a farm, and other structures provide a look into the past.

5. Home tours: Located in the heart of Batsto Village, a 32-room mansion with Italianate architecture once housed the town's ironmasters. Joseph Wharton, a rich Philadelphian who owned the home for numerous years, later refurbished it. Today, you can see 14 rooms of the mansion with a guide; however, tours are only offered a few days a week, so check the Batsto website for specifics.

6. Outdoor activities:Hiking, kayaking, canoeing, biking, bird viewing, and wildlife spotting are just a few of the outdoor activities available. Some people like the peace, but others believe the area's deep, remote forests are home to the famous "Jersey Devil." The Batsto Lake route is one of the more popular trails (easy, marked, and accessible). The Mullica River trail and the Batona trail are two more that are moderate to challenging.

What Is The Best Way To Get To Batsto Village?

Batsto Village is open all year, and because New Jersey has four seasons, it will be different each time you visit. Although there is much to do in the summer, if you don't mind the cold, it may be a particularly beautiful place after a winter snowfall. If you take a tour, hike, or visit the historic sites, you may expect to spend a few hours here.

You can go canoeing, mountain biking, or strolling around the nature trails on your while at Batsto. Visit the Visitor's Center to discover more about the tours available about the town's history. Batsto Village also provides a variety of group activities, such as monthly treks, astronomy tours, and special events for kids. Daily self-guided tours begin at the Visitor's Center and last about an hour. On some days, guided mansion tours are available.

Nearby Attractions

There aren't many locations within a short drive because this is a pretty remote and quiet part of Burlington County. Hammonton, about 10 miles distant, is the closest town. If you have more time, it's worth stopping by the centre of town, where there are various casual cafes and boutiques.

Conclusion

Forest fires are usually only seen during certain times of the year. They normally start during the dry season and can be avoided if you take precautions. Forest-fighting funding has been granted in previous Five-Year Plans. During British time, summer fires were prevented by clearing forest trash all along the forest boundary. "Forest Fire Line" was the name of the project. This line was employed to keep the fire from spreading from one compartment to the next. The litter was gathered and disposed of in a different area.

The fire will only expand if there is a constant supply of fuel (dry grass) along its route. The simplest strategy to control a forest fire is to keep it from spreading, which can be accomplished by constructing firebreaks in the form of shallow ditches in the woods.

Precautions

The following are crucial fire safety precautions:

- To keep the fire or ignition source away from combustible and non-combustible materials.

- To keep an eye on the source of the fire and exert control over it.

- Allowing combustible or flammable materials to pile up excessively and stocking them according to the protocol for safe storage of such combustible or flammable materials is prohibited.

- Places near forests, such as factories, coal mines, oil stores, chemical facilities, and even domestic kitchens should adopt safe procedures.

- Incorporating fire-fighting and fire-prevention tactics and equipment.